Economic costs of healthcare-associated infections acquired in an Intensive Care Unit

2024

To José and Alda, my dear parents.

To Fabíola, my beloved wife and companion.

To Fernanda, José, Sarah and Oscar, my continuation.

To Lucas, my dear stepson.

The very first requirement in a hospital is that it should do the sick no harm".

Florence Nightingale (nurse in the Crimea war, 1820 – 1910)

SUMMARY

LIST OF FIGURES

LIST OF ABBREVIATIONS AND/OR ACRONYMS

AMIB	Associação de Medicina Intensiva Brasileira / Brazilian Intensive Care Medicine Association
ANVISA	Agência Nacional de Vigilância Sanitária / National Health Surveillance Agency
APACHE	Acute Physiology and Chronic Health Evaluation
APIC	Association for Professionals in Infection Control and Epidemiology
ASP	Antimicrobial Stewardship Program
BAL	Bronchoalveolar Lavage
BSI	Bloodstream Infection
CBIC	Certification Board of Infection Control and Epidemiology
CDC	Centers for Diseases Control and Prevention
CDI	*Clostridioides difficile* Infection
CFM	Conselho Federal de Medicina / Federal Council of Medicine
CFU	Colony-Forming Units
CNES	Cadastro Nacional de Estabelecimentos de Saúde / National Registry of Health Establishments
CNS	Coagulase-Negative *Staphylococcus*
CREMESP	Conselho Regional de Medicina de São Paulo/ Regional Council of Medicine of São Paulo
CVC	Central Venous Catheter
ESBL	Extended-Spectrum Beta-Lactamase
ETA	Endotracheal Aspirate
FAPIC	Fellow of the Association for Professionals in Infection Control and Epidemiology
FORMSUS	Formulário de notificação via web / Web notification form
HAI	Healthcare-Associated Infection
HI	Hospital Infection
HICS	Hospital Infection Control Service
IBC	Indwelling Bladder Catheter

ICU	Intensive Care Unit
iOS	iPhone Operating System
IP	Infection Preventionist
LOS	Length of Stay
MBA	Master Business Administration
MBP	Mean Blood Pressure
MDR	Multidrug Resistance
MRSA	Methicillin-resistant *Staphylococcus aureus*
MSSA	Methicillin-sensible *Staphylococcus aureus*
NNISS	National Nosocomial Infection Surveillance System
OHC	Oral Hygiene with Chlorhexidine
OPM	Orthoses, Prostheses and Materials
OTT	Orotracheal Tube
PAMV	Pneumonia Associated with Mechanical Ventilation
PEN	Parenteral Nutrition
R$	Brazilian Real
RA	Rest Room
SENIC	Study on the Efficacy of Nosocomial Infection Control
SIGTAP	Sistema de Gerenciamento da Tabela de Procedimentos, Medicamentos e OPM / Medications and OPM Table Management System
SSI	Surgical Site Infection
SUS	Sistema Único de Saúde / Health Unique System
US$	American Dolar
USA	United States of America
UTI	Urinary Tract Infection
MV	Mechanical Ventilation
WHO	World Health Organization

ABSTRACT

Introduction: Healthcare-Associated Infections (HAI) are 5-to-10-fold higher incidence in an Intensive Care Unit (ICU) in comparison to the rest of a hospital. The literature rates vary from 10.0% to 66.2%. HAIs are an important cause of mortality, transient or permanent functional disabilities, increased length of ICU stay and costs of a treatment.

Objective: Evaluating the magnitude of patient economic costs for patients who acquired HAI during their hospitalizations in an ICU.

Methods: Exploratory, retrospective study of patients who were admitted to the ICU of a public hospital through January to March 2018.

Results: Seventy-eight patients were admitted to the ICU. The hospitalization mean time was 26.4 days (patients with HAI acquired in the ICU), 8.4 days (patients already hospitalized with infection), and 6.4 days (patients without infection). The hospitalization costs were respectively 38.4%, 35.7% and 25.9% of the total costs. Thirteen patients with multidrug-resistant germs (MDR) averaged 25.5 days hospitalized, and the five patients without MDR infection, 4.4 days.

Discussion: In addition to HAI acquired in the ICU, other predictors of prolonged stay, and consequently higher treatment costs, were the use of mechanical ventilation and infection with MDR microorganisms.

Conclusion: HAI are associated with longer hospital stay and higher costs for treatment.

Key words: Healthcare-Associated Infections; Intensive Care Unit; costs.

PREFACE

In this preface I would like to introduce the reader to the Association of Infection Control Professionals (APIC). APIC is an organization focused on the development of professionals dedicated to infection control in the most diverse environments where healthcare occurs. For example, outpatient clinics, infirmaries, emergency rooms, and especially Intensive Care Units (ICU).

The ICU, as we will see later, is one of the most favorable hospital environments for the emergence of Healthcare-Associated Infections, often with multidrug-resistant germs.

The *Association of Professionals for Infection Control* (APIC) presents us with a professional development system. This system, called *APIC's Competency Model* for the Infection Preventionist, outlines a strategy for the professional development of infection preventionists. First, the APIC competency model identifies which fields of knowledge, skills and attitudes IPs must possess: research, leadership, professional stewardship, quality improvement, IPC operations and IPC informatics (figure 1).

Each of these fields of knowledge, skills and attitudes, on the other hand, has specifications to be achieved by infection control professionals or infection preventionists (IP) (see table below).

Figure 1 – APIC's Competency Model

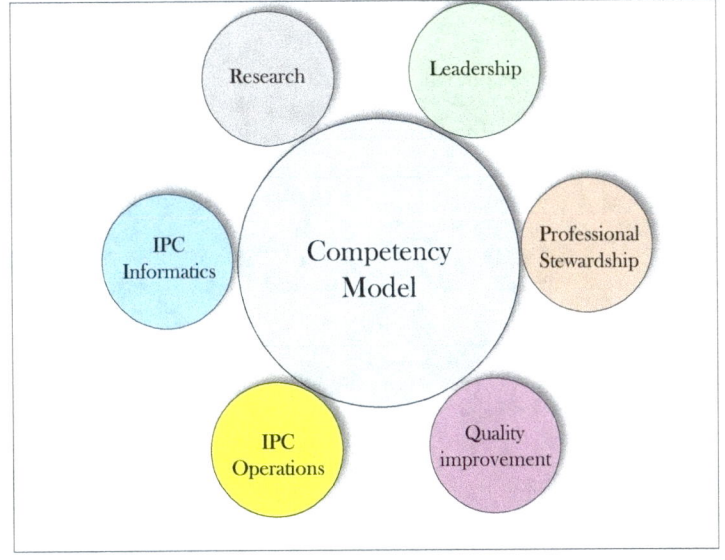

Adapted from Am J of Infect Control, v. 47, n. 6, p. 602-14, 2019

As we can see in table below, IPs must have diverse knowledge. But I would like to highlight the importance of *cost management*, the theme of this book.

The IPs can be evaluated for financial issues through the *program management subdomain* of the **Leadership competency** and the *financial acumen subdomain* of the **Professional Stewardship competency**. Billings C, Bernard H, Caffery L, et al. (2019) refer that by demonstrating skills management of budgets, resources, personnel and programs, IPs will increase their credibility with stakeholders during decision-making. Furthermore, IPCs must project the impact of IPC activities across the facility, quantifying the costs and savings of IPC initiatives (e.g.

implementation of new programs, new practices or new products).

Fields of knowledge, skills and attitudes / specifications

Leadership	Professional stewardship	Quality improvement	IPC operations	IPC informatics	Research
Communication	Accountability	IP as subject matter expert	Epidemiology and surveillance	Surveillance technology	Evaluation of research
Critical thinking	Ethics	Risk assessment and risk reduction	Outbreak detection and management	EMR and electronic data warehouse	Conduct or participate in research or evidence-based practice
Collaboration	Financial acumen	Data utilization	Diagnostic stewardship	Data management, analysis and visualization	Implementation and dissemination science
Behavioral science	Population health	Performance improvement	Education		Comparative effectiveness research
Program management	Continuum of care	Patient safety	IPC rounding		
Mentorship	Advocacy		Cleaning, disinfection, sterilization	Application of diagnostic testing data and techniques	
			Antimicrobial stewardship		
			Emerging technologies		

Adapted from Am J of Infect Control, v. 47, n. 6, p. 602-14, 2019.

Another aspect that I would like to mention is that, as the infection preventionist progresses in his knowledge, he has the possibility of reaching higher levels within the APIC's Competency Model (see figure 2).

Figure 2 – APIC's five-level Competency Model for IP

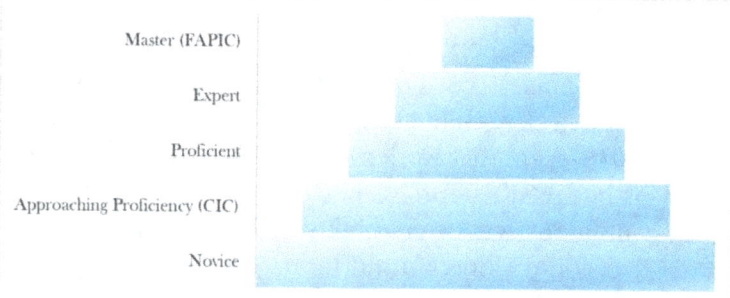

Master (FAPIC)

Expert

Proficient

Approaching Proficiency (CIC)

Novice

Adapted from Am J of Infect Control, v. 46, n. 8, p. 921-7, 2018.

An instrument used by APIC to assess the level of competence of various IPs around the world, is the Certification Board of Infection Control and Epidemiology (BERNARD H, HACKBARTH D, OLMSTED RN, et al, 2018; HENMAN LJ, CORRIGAN R, CARRICO R, et al, 2015). And to reach the Master level, the IP must complete a Fellowship Program in this area, also known as FAPIC.

This is the monograph I presented in 2019 in the **MBA** in Health Management and Infection Control[1]. I thought it would be a waste to keep a copy of it on my shelf and not share it with other people who are also interested in the subject.

I chose this subject among the many others that exist - in the field of infection control - because in addition to being a medical doctor, I have a degree in Hospital Administration. So, I made a link between my two fields of knowledge.

I must emphasize that at the time I wrote this monograph, I still did not know APIC. Today I am an associate member and am certified by the Board of Infection Control and Epidemiology (CBIC).

March 3rd, 2024.
The author.

---

[1] MBA CCIH Cursos, held in Rio de Janeiro.

INTRODUCTION

Hospital or nosocomial infections have been known since the emergence of hospitals. The incidence of hospital-acquired infections was high for a long time, due to the high prevalence of epidemic diseases and the precarious hygiene conditions that existed until then. However, it was only in the first half of the 19th century that these infections began to be discussed by health professionals, with Ignaz Semmelweis, Florence Nightingale, James Young Simpson and Joseph Lister being the greatest exponents of this century (MARTINS, 2001a).

HAIs gained importance during the 20th century, mainly due to the increase in advanced life support, the use of immunosuppressive therapies, the increase in population longevity and the increase in resistance to antimicrobials (PADOVEZE, 2014). Furthermore, according to this author, the term hospital infections (HI) was replaced in the 1990s by healthcare-related infections, which is more appropriate, as these infections can also be acquired in other environments, such as offices and homes (home care).

In general, healthcare-associated infections (HAIs) in hospitals can be defined as "any clinical manifestation of infection that presents itself within 72 hours of admission", when there is no clinical or laboratory evidence of this morbidity. at the time of admission. HAIs can also manifest themselves after hospital discharge, in this case when they can be related to hospitalization or hospital procedures

(BRASIL, 1998; JEON et al., 2012; MARTINS, 2001b). The Centers for Diseases Control and Prevention (CDC) (apud TIWARI; ROHIT, 2013; PRADHAN; BHAT; GHADAGE, 2014) also defines nosocomial infection as a [...]

> [...] localized or systemic condition resulting from an adverse reaction to the presence of an infectious agent or toxins, without any evidence that the infection was present or incubating at the time of admission to the emergency department.

The probability of contracting an HAI in the ICU is 5-10 times higher than in the rest of the hospital and may represent 20% of all infections that occur in a hospital institution (LIMA; ANDRADE; HAAS, 2007).

North American data indicate that HAIs constitute the fifth cause of mortality in hospitals that care for acute patients. Approximately two million patients are affected by these infections annually in the United States of America (USA), of which around ninety thousand (4.5%) die. It was estimated that the annual hospital costs in the USA involved with diagnosing, treating and preventing these infections were US$ 28-45 billion. Furthermore, according to the author of the work, the main infections acquired within the hospital are bloodstream infections (BSI), surgical site infections (SSI), urinary tract infections (UTI) associated with the use of a bladder catheter and ventilator-associated pneumonia (VAP) (STONE, 2009).

In addition to direct institutional costs, HAIs can cause temporary or permanent functional disabilities, emotional stress and prolonged hospital stays, which, in turn, also cause additional costs for patients (TIWARI; ROHIT, 2013).

Regarding Intensive Care Units (ICU), in a study also carried out in the USA, HAIs occurred in around 10% of all patients admitted to these units, at a rate of 21-25 infections/1,000 patient-days, and with a mortality rate of 10%. Of this total, 27% were pneumonia (86% of which were associated with mechanical ventilation), 31% were UTI (95% of which were associated with the use of a urinary catheter), and 19% were primary bloodstream infections (87% of which were associated with the use of central catheters) (JAIN et al., 2006). A national, prospective study found an incidence of 66.2% among 71 patients, with a mortality rate of 57.5% among those who acquired the infection in the ICU (LIMA; ANDRADE; HAAS, 2007).

In addition to the predisposing factors for ICU-acquired HAIs related to invasive procedures, other factors related to patients and infection control can also facilitate the onset of these infections (BEEKMANN; HENDERSON, 2005; STRAUSBAUGH; 2005; WARREN, 2005). Some of the risk factors cited by these authors are severity of the underlying disease, immunodeficiency, loss of skin integrity, advanced age, reduced level of consciousness or coma, metabolic acidosis, chronic lung disease, diabetes mellitus, altered renal function, disability in hand hygiene measures.

A tool used both for prognostic purposes and to calculate the risk of an ICU patient acquiring a nosocomial infection is the APACHE II scale (SUKA; YOSHIDA; TAKEZAWA, 2004). According to these authors, this scale only assesses intrinsic risk factors (the patient's own), it is not possible to assess extrinsic risks (invasive interventions).

While there is a lot of information about the costs of nosocomial infections in developed countries, the same does not occur in developing countries. There are three reasons for not making a cost inference from data from developed countries: different microbiological profiles; doctors' reluctance regarding treatment, based on the impression that survival is low; and limitation of resources allocated to the treatment of these infections, prioritizing other patients with prospects for better evolution (CHACKO et al, 2017).

Hospital-level infection surveillance, prevention and control programs date back to the 1950s, having been refined during the 1960s and 1970s. With the objectives of measuring the progression of adoption of infection control programs by US hospitals, and to determine the degree of reduction in IH rates through the implementation of these programs, the Study of the Efficacy of Control of Nosocomial Infections (SENIC) was carried out during the 1970s. The study, which took place during the 1970s, found that IH rates were reduced by approximately 32% if control and surveillance programs had included four components: appropriate surveillance activities and vigorous control efforts; at least one full-time infection controller for every 250 beds; a trained hospital epidemiologist; and feedback to surgeons on surgical infection rates. Considering the results of the reduction in the incidence and costs related to hospital infections, the SENIC study concluded that infection control programs were effective. In 1985, while the annual cost of IH was estimated at US$ 4 billion, the annual cost of maintaining infection control programs was estimated at US$ 60,000 for every 250 beds, with a projection of US$ 243 million for throughout the country (PUGLIESE, 1992).

To standardize the diagnosis and epidemiological surveillance of HAIs, Anvisa[2] (2017) published the manual of Diagnostic Criteria for Infections related to Health Care, which contains the criteria and flowcharts that allow case identification, collection and interpretation and the dissemination of information in a systematic way by professionals and managers of the health system.

Currently, healthcare-related infections and their control are on the agenda of several North American entities focused on quality of care (Joint Commission on Accreditation of Healthcare Organizations), for hospital administrators (American Hospital Association), for health professionals (Association for Practitioners in Infection Control, Society for Hospital Epidemiology of America,

---

[2] Anvisa is "Sanitary Vigilance National Agency".

Surgical Infection Society) and public health (Public Health Service).

In Brazil, there is still a lack of data regarding the economic impacts of infections acquired in the ICU. Isolated data from an institutional survey carried out in a philanthropic hospital shows a longer length of stay and greater daily spending on ICU stays (NANGINO et al.; 2014).

With an increased length of stay due to HAIs acquired in the ICU, prevention measures are of great relevance, especially in Brazil, where there are only around 45,000 ICU beds (Conselho Federal de Medicina[3], 2018), distributed among 24% of hospitals in the country (National Registry of Health Establishments, apud Associação de Medicina Intensiva Brasileira[4], 2016). AMIB (apud CFM, 2018) suggests that the ideal quantity is 1-3 beds/10,000 inhabitants. This deficit of ICU beds requires bed management in this sector of Brazilian hospitals to be extremely efficient, even more so due to the peculiarities of this sector of a hospital. According to CREMESP, Regional Council of Medicine of the State of São Paulo (apud OLIVEIRA et al., 2010),

> [...] intensive care unit (ICU) is the hospital room intended for the care of serious or at-risk patients, potentially recoverable, who require uninterrupted medical assistance, with the support of a

---

[3] Conselho Federal de Medicina is "Federal Council of Medicine".

[4] Associação de Medicina Intensiva Brasileira is the "Brazilian Intensive Medicine Association".

multidisciplinary health team and other specialized human resources, in addition to equipment.

The hospital, from which the casuistry of this monograph was collected, is a medium-sized, tertiary public hospital, located in the State of Rio de Janeiro, with 70 beds distributed as follows: ICU (09), wards (41), rest (20). It has a Surgical Center, Emergency Room, Pharmacy, Clinical Analysis Laboratory, Laundry, Warehouse, Kitchen, Coffee shop, Medical Records Archive, Blood Bank, Hospital Infection Control Service (HICS). The ICU has an isolation bed and only serves adult patients.

Figure 03 shows the patient flows between the hospital sectors, with emphasis on the ICU. Patients can be admitted to the ICU:
a) directly from the community, after care in the Emergency Room,
b) from the wards,
c) from the Adult Rest Room,
d) from the Yellow or Red Rooms (located in the Emergency Room) or,
e) from another hospital, upon prior request for a vacancy. Upon discharge from the ICU, the patient is generally transferred to the ward or may also return to the hospital of origin.

Pediatric patients are treated at the Children's Hospital, attached to the hospital under study, and those in serious condition are referred to hospitals affiliated with the public network.

It is important to highlight that the present study is justified by the still small number of publications in developing countries about the economic impact of HAIs acquired in ICUs.

OBJECTIVES

Main objective:

To evaluate the magnitude of the economic costs of patients who acquired healthcare-associated infections (HAIs) during their stay in an Intensive Care Unit.

Secondary objectives:

Compare the costs of patients with HAIs acquired in the ICU, with the costs of patients also admitted to this unit, but with infections acquired outside the ICU, and with the costs of patients without infection.

Verify whether nosocomial infections acquired in the ICU are associated with an increase in the length of stay and, consequently, in costs.

Check whether infections with MDR microorganisms are associated with an increase in hospitalization time and, consequently, costs.

METHODS

Delimitation of the study object

The economic costs of hospitalizations of patients who acquired HAIs in the ICU constitute the scope of this study. Clinical and microbiological aspects will also be evaluated to verify the influence of these aspects on the length of stay and, consequently, on the final costs of hospitalization.

Data collecting

Data relating to patients were extracted from epidemiological surveillance forms, located at the Hospital Epidemiology and Infection Control Service. A model of the Epidemiological Surveillance Form can be found in the annexes of this monograph.

Regarding the diagnostic and therapeutic procedures carried out, the nursing shift record books were searched. The surgeries performed during the study period were recorded in the Surgical Center's monthly reports. To check the values for procedures and surgeries, the SUS Procedures, Medications and OPM Table Management System (SIGTAP), available on the web, was consulted.

Medical records were consulted to verify the APACHE II score.

To verify the number of patient days, the rates of use of invasive devices and the monthly numbers of

HAIs, the notification forms for FormSus[5] were consulted.

Project characterization

This is an exploratory, retrospective study of patients who were admitted to the ICU of a public hospital. The results found will be compared with those in the literature on the subject.

Definition related to population and sample

Initially, all patients admitted to the ICU were considered, and who were monitored by the hospital's epidemiological surveillance activities during the months of January, February and March 2018. 44 patients who remained hospitalized for a period of less than three days (or 72 hours) were excluded. . For it to be considered as having been acquired in the ICU environment, HAI would need to manifest itself clinically and/or laboratory after this time interval (BRASIL, 1998; MARTINS, 2001b; GLANCE et al., 2011). If a patient had been admitted for treatment of an infectious condition and, during their stay in the ICU, had developed another infection, this patient was classified as being in the group of patients with HAIs acquired in the ICU.

The planning for the execution of the project considered the ease of collecting data and the limited period to carry it out. The research of medical records, epidemiological surveillance forms, ICU and

---

[5] FormSus is "electronic form available in web".

Surgical Center record books, data tabulation and bibliographic research were carried out by a single person. Regarding the survey of costs, we chose to research the primary sources mentioned above.

For the diagnosis of Surgical Site Infection (SSI), Bloodstream Infection (BSI) and Healthcare-related Pneumonia, when associated with mechanical ventilation (VAP), the diagnostic criteria presented in figures 13, 14 and 15 A/B were followed. respectively. These tables are integral parts of the publication "Diagnostic Criteria for Healthcare-Related Infection" (BRASIL, Agência Nacional de Vigilância Sanitária, 2017).

Endotracheal aspirate (ETA) was used as a respiratory secretion collection technique for microbiological diagnosis of VAP and cultures were quantitative, considered positive when there were ≥ 105 CFU/mL.

All samples collected for microbiology examination purposes were sent to an approved private laboratory.

To calculate the APACHE II scale score, the MedCalc app version 2.7.5 for IOS[©] 2008-2013 P. Pfiffner & M. Tschopp was used.

Data analysis and interpretation

The Preventionist doctor, according to the information collected through the epidemiological

Surveillance Bulletin (figures 16 and 17), classified the patients as "not carrying infection", "carrying infection acquired outside the ICU" and "carrying HAI acquired during stay in the ICU". If a patient was admitted to the ICU more than once during their hospital stay, they were considered a new patient.

The economic costs of hospitalizations for patients who acquired HAIs during their stay in the ICU will be compared with the same types of costs for patients who already had infectious conditions before admission to the ICU and patients who did not have infections during their stay in the ICU.

RESULTS

Casuistry

One hundred and twenty-two patients were admitted to the ICU from January 1st to March 31st, 2018. Forty-four patients, who remained hospitalized for less than 72 hours (three days), were not included in this series for study purposes, as outlined in the study methodology.

The age of the patients varied between 16 and 83, with an average of 58.0 years. ICU admissions were clinical and surgical in nature, as shown below. Analyzing the Epidemiological Surveillance Records, it was not possible to define the underlying cause for the hospitalization of two patients.

Clinical nature	n	Surgical nature	n
Infectious cause	30	Postoperative period	7
Cardiovascular cause	23	Gunshot injury	4
Pancreatitis	2	Polytrauma	3
Acute respiratory failure	1	Traumatic brain injury	3
Acute renal failure	1	Perforated ulcer	1
Exogenous intoxication	1	-	-
Total	58		18

The number of patient days was 218 in January, 223 in February and 215 in March.

Figure 04 shows the distribution of patients according to the presence or absence of infection and, if an infection occurred during hospitalization in the ICU, whether it had been acquired before or after admission to this sector.

Length of stay in the ICU

The twelve patients who developed some type of HAI in the ICU were hospitalized between five and 45 days, with an average of 26.4 days (figure 05).

For the 33 patients who already had an infectious condition before admission to the ICU, the length of stay was shorter, from four to 27 days, with an average of 8.4 days (figure 05).

As for the 33 patients who did not present an infectious condition during their ICU stay, the LOS was the shortest of the three groups, ranging from four to 17 days, with an average of 6.4 days (figure 05).

Considering the daily decrease in the number of patients admitted over the days, this decrease was more pronounced for patients without infection, followed by that of patients who were already admitted to the ICU with infection, and the slower decrease occurred among patients with HAIs acquired in the ICU (figure 06).

The most frequently performed invasive procedures during the ICU stay

Considering the three months of the series, among all 78 patients admitted to the ICU for a period longer than 72 hours, urinary catheterization was present in (76.9%), followed by peripheral (57.7%) or central venous access (56.4%), and orotracheal intubation or tracheostomy (55.1%). Less frequently there were patients with gastric, enteral and dialysis catheterizations; parenteral nutrition; chest drain; and intravascular access to measure mean arterial pressure (MAP) (figure 07).

Regarding the rates of use of invasive devices, the following were notified on FormSus:

	January	February	March
Central venous catheter-day	116	150	150
Indwelling bladder catheter-day	155	188	158
Mechanical ventilation-day	126	101	94

Topographies of infectious processes among the 78 patients in the study

The graph in figure 08 shows that pneumonia was the most common infection, whether acquired before or after admission to the ICU. Less frequently, bloodstream, urinary tract, intra-abdominal, surgical site, and skin/soft tissue infections occurred.

As shown in the previous topic, although bladder catheterization was present more frequently in the period considered, only one patient had an episode of UTI in the ICU.

Regarding the monthly numbers of HAIs acquired in the ICU, the following were notified through FormSus:

	January	February	March
BSI associated with central venous catheter	-	1	-
UTI related to indwelling urinary catheter	-	1	-
Pneumonia associated with Mechanical Ventilation	3	4	3

Results of microbiological tests carried out during patients' stay in the ICU

From January to March 2018, 106 clinical specimens (blood, endotracheal aspirate, urine) were collected from patients admitted to the ICU, for microbiological culture purposes, as shown below:

	blood		endotracheal aspirate		urine		total	
	pos	neg	pos	neg	pos	neg	pos	neg
Jan	5	10	18	3	1	11	24	24
Feb	7	10	12	3	-	6	19	19
Mar	7	6	2	1	-	4	9	11
Total	19	26	32	7	1	21	52	54

The distribution of microorganisms, according to the culture results of the clinical specimens collected (endotracheal aspirates, blood and urine) is shown below:

	Tracheal secretion	Blood	Urine	Total
Staphylococcus aureus	6	5	-	11
Klebsiella pneumoniae	5	3	1	9
Enterobacter sp.	3	2	-	5
Pseudomonas aeruginosa	5	-	-	5
Acinetobacter sp.	4	-	-	4
Estafilococos coagulase-negativo	-	2	-	2
Stenotrophomonas maltophila	2	-	-	2
Morganella morganii	-	2	-	2
Escherichia coli	1	1	-	2
Proteus mirabillis	2	-	-	2
Haemophilus influenzae	1	-	-	1
Burkholderia cepacia	1	-	-	1
Providencia sp.	1	-	-	1
Total	31	15	1	47

Figures 09 and 10 show, respectively, the distribution of these microorganisms among patients with HAIs acquired in the ICU and among patients who already had an infectious condition upon admission to the ICU.

The thirteen patients who presented at least one MDR germ (figure 11) in their cultures had an average stay of 25.5 days in the ICU, with a range of five to 45 days. On the other hand, the five patients who did not present MDR microorganisms in their cultures had an average stay of 4.4 days in the ICU, with a variation of three to seven days.

The APACHE II score at patient admission

It was possible to evaluate the APACHE II scale of 54 patients, as shown below:

APACHE II (score)	n	deaths	% deaths
0 – 5	6	0	0,0%
6 – 10	7	2	28,5%
11 – 15	13	5	38,4%
16 – 20	10	5	50,0%
21 – 25	5	2	40,0%
26 – 30	10	8	80,0%
> 30	3	2	66,6%
Total	54	24	44,4%

Four medical records were not found and in twenty records the APACHE II results were not expressed because there was a lack of data to calculate it (complementary diagnostic tests or Glascow scale).

With a median APACHE II score between 16 and 17, the following results were obtained:

APACHE II	HAI in the ICU	Infection prior to admission	total
0-16	3	24	27
17-41	7	20	27
total	10	44	54

The table above shows that there was a direct relationship between the APACHE II score and the number of infections after admission to the ICU. The same did not occur among patients who did not develop HAIs in the ICU.

The costs of hospitalizations

The costs of ICU admissions at the hospital in this study basically have three components: the hospital daily rate; the diagnostic/therapeutic procedures and surgeries performed; and physiotherapy sessions.

The daily rate has a fixed value (R$ 478,72), regardless of the use of mechanical ventilation, medications (including antimicrobials) or disposable material, and was the most relevant cost component. On the other hand, diagnostic and therapeutic procedures (for example, computed tomography, radiography, echocardiography, Doppler ultrasound of the lower limbs, endoscopies, obstetric ultrasound, hemodialysis and surgeries) and physiotherapy corresponded to a smaller share of costs.

According to the costs generated, the distribution of the three groups of patients was as follows:

	Daily (R$)	Procedures and surgeries (R$)	Physiotherapy (R$)	Total (R$)
No infections	102.445,84	6.570,00	565,07	109.580,91
Infections outside the ICU	138.349,36	12.405,94	462,33	151.217,63
IRAS in the ICU	151.754,24	10.531,03	387,61	162.672,88
Total (R$)	392.549,44	29.506,97	1.415,01	423.471,42

Figure 12 shows the average costs for each of the three groups (HAIs in the ICU, infections acquired outside the ICU, and without infection) with each component of hospitalization costs (hospital daily rates, diagnostic and therapeutic procedures, surgeries, physiotherapy). To construct this graph, the group of patients with HAIs acquired in the ICU was considered as a reference for 100% of average expenses.

DISCUSSION

Hospital infections are the most frequent complication in hospitalized patients (BRAGA, 2015; GLANCE et al., 2011). Braga (2015) reports an estimated incidence of 5 to 15%, Pradhan, Bhat and Ghadage (2014) found 9.6% in their study, while Glance et al. (2011) cite an incidence of 4.5 IH for every hundred admissions to North American hospitals. The rate of infection acquired in the ICU in the present case series was 15.4%, being at a similar level to that of other authors.

Within this context, healthcare-related infections can greatly affect patient safety and, therefore, the quality of care. In Braga's words (id., p. 16),

> [...] This is an often-avoidable adverse event that can lead to increased mortality, definitive and transitory sequelae, prolongs the length of hospital stay and requires diagnostic and therapeutic interventions that generate additional costs to those already determined by the underlying disease of the patient.

Several authors have correlated HAIs with increased length of stay in hospital, increased costs and even increased morbidity and mortality. Infection acquired within the ICU, according to Olaechea et al. (2003) is significantly associated with the length of stay, which is not verified with other variables related to the severity of the disease, assessed by APACHE II, emergency surgeries, nor with infections acquired before admission to this treatment unit.

Analyzing from the economic perspective of a prolonged stay in the ICU, Kerlin and Cooke (2015) emphasize that simply reducing this length of stay is an insufficient measure to save money. According to these authors, the fixed costs of an ICU correspond to 80% of total costs and, therefore, using length of stay (which varies depending on the patient) as the main cost measure would be erroneous. Graves (2004) also cites similar values (84-89%) for hospital fixed costs.

TAN et al. (2012) present a standardized cost methodology, which is more useful for comparing costs between hospital institutions. They classified costs into four components: hospitality and nutrition; diagnostics; consumables and work (of doctors and nurses). In the seven ICUs studied, in four European countries, it was found that diagnostic costs (examinations) were 14%, and consumable costs (medicines and blood products) were 22%; with hospitality/nutrition they were 4%; and with the work of professionals, it was 60%. In a way, these data are in line with those of the authors who classify costs as fixed and variable, as hospitality, nutrition and professional work are fixed and correspond to most costs in an ICU.

Scott (2009) also makes a distinction between spending on goods and services that the hospital offers, and spending on each patient, making those expenses a more accurate estimate of the economic value of the resources used during hospital care.

In a different way, Nangino et al. (2012) carried out a study of the financial impact of HAIs in the ICU, considering only the costs of medicines and materials, comparing patients with and without infection, and with adjustment (correction factor) for the period of stay in the ICU. They did not specify whether the infection was acquired in the ICU or not. However, they cite the increase in length of stay as an "important limitation of access to intensive care".

Another study, also a case-control study, evaluated the costs of medicines, consumables, hospital stays, diagnostic tests and procedures, blood products, staff visits and antimicrobials. The patients' main expenditure was on medicines, with antimicrobials accounting for almost ½ of the value of this item consumed (TIWARI; ROHIT, 2013). Antibiotics were also responsible for more than half of medication costs in the ICU, according to Biswall et al. (2006 apud JAYARAM; RAMAKRISHNAN, 2008).

The use of severity/mortality risk scores can reduce the number of hospitalizations of low-risk patients, avoiding unnecessary expenses and optimizing costs. Patients identified as being at low risk of mortality could benefit from care at a lower level of complexity (CZAJKA; SALYER, 2017; THOMAS et al., 2015).

According to Chacko et al. (2017), an infection acquired in the ICU can double the costs compared to those patients who do not develop an infection. This information refers, once again, to the importance of not admitting patients with low clinical severity to an ICU, sparing them from an adverse and unwanted event such as an HAI.

Stone (2009) reminds that when determining the costs attributable to nosocomial infections, the severity of the underlying disease, comorbidities, as well as the length of stay in hospital before acquiring the infection are considered.

Angelis et al. (2010) point out that the costs associated with hospital-acquired infections (HI), if calculated after hospital discharge, may be underestimated. These authors tell us that if the time prior to infection is not considered, for the purpose of matching with patients without infection, the length of hospital stay attributable to HI can be considerably reduced. In this way, the costs corresponding to HI become more "concentrated".

Three studies considered all infection topographies when assessing costs. The study by Glied et al. (2016) was based on three hospitals, between 2006 and 2012, comparing patients with infection with control patients without infection. The Zimlichman et al. (2013) was based on a meta-analysis in North American hospitals between July 2011 and April 2013, considering the costs of each type of infection and the overall costs of each type of infection in the USA. Tiwari and Rohit (2013) also carried out a case-control study, but in a single private hospital in India. Comparing the results of the three studies mentioned, the nosocomial infections with the highest costs were, in descending order:

GLIED et al. (2016)	ZIMLICHMAN et al. (2013)	TIWARI; ROHIT (2013)
Pneumonia	BSI	VAP
BSI	VAP	BSI
UTI and SSI	SSI	ITU
-	ICD	SSI
-	catheter-based UTI	-

Considering the group of people, the nosocomial infections that produced the highest costs were in descending order: surgical site infections (SSI), VAP, BSI, *Clostridioides difficile* infections (CDI) and UTI, as shown in figure 18 (ZIMLICHMAN et al., op. cit.).

Other authors restricted themselves to certain types of nosocomial infections to base their work on costs: ventilation-associated pneumonia (ALP et al., 2012; ERBAY et al., 2004; KOLLEF; HAMILTON; ERNST, 2012; MATHAI et al., 2015; NIEDERMAN, 2001; SAFDAR et al., 2005), primary bloodstream infections (DiGIOVINE et al., 1999) and infections in polytrauma patients (GLANCE et al, 2011).

The large number of studies on nosocomial pneumonia, especially VAP, may be because it is the most common infection in almost all ICUs (SWANSON; WELLS, 2013), as well as the one with the highest mortality (ERBAY et al., 2004) among the other HAIs. In the present study, VAP was the most common ICU-acquired infection, corresponding to 62.5% (10/16) of all HAIs (figure 08).

Erbay et al. (id., 2004) carried out a case-control study in which they found that respiratory failure, length of stay, score less than 9 on the Glascow scale at the time of admission, and enteral feeding were risk factors for acquiring VAP, adding on average 5.9 days of stay and an excess cost of US$5,683.00.

The literature shows that VAP increases the average length of hospital stay by 6.1 (SAFDAR et al., 2005) up to ten days (KOLLEF et al., 2012), as well as generating an additional cost of US$ 13,647.00 with diagnostic tests and use of antimicrobials (SAFDAR et al., op. cit.) up to around US$ 40,000.00 for personnel, pharmacy, ventilation, respiratory therapy and x-rays (KOLLEF et al., op. cit.).

Niederman (2001), in his literature review study on the cost-effectiveness of VAP treatment, refers to the costs due to resistant microorganisms, diagnostic tests, therapy and prevention of these infections.

Resistant microorganisms, which will be discussed later, can increase costs because of "inadequate" initial empirical therapy (NIEDERMAN, 2001; SWANSON; WELLS, 2013) and due to the requirement for isolation beds (NIEDERMAN, 2001). The use of more expensive invasive techniques such as fiberoptic bronchoscopy increases diagnostic costs. The combined use of more expensive antimicrobials, for a longer period, and the treatment of complications resulting from the infection increase treatment costs. Prevention measures, which require training and dedication from the ICU team to patient care, also increase hospital costs (NIEDERMAN, 2001).

Studies on VAP carried out in developing countries have also been very frequent. Alp et. al. (2012) found that VAP is responsible for a four-fold increase in the length of stay in the ICU, at a cost three times greater, compared to patients without mechanical ventilation. Mathai et al. (2014) mention that the main components of the costs were medications and the increase in the length of stay in the ICU, corroborating the findings of other studies.

The clinical diagnosis of VAP has high sensitivity (NIEDERMAN, 2001), but low precision (CORRÊA et al., 2014), which has led to the use of quantitative cultures of respiratory secretion samples to confirm the diagnosis. According to Swanson and Wells (2013), only around 40% of patients with clinical suspicion will have VAP confirmed. For Corrêa et al. (id., 2014), the endotracheal aspirate (ETA) technique with quantitative culture is viable for diagnosis, with no worse results compared to other techniques, in terms of mortality, length of stay in the ICU, and proportion of patients who respond to treatment. Furthermore, ETA is a simple, non-invasive, easy to perform and low-cost method (McCAULEY et al., 2016; SHIN et al., 2011). Carvalho et al. (2004) highlight the need to use cost-effective and rapid methods to avoid delaying antimicrobial therapy while awaiting BAL. In the study they carried out, a sensitivity of 72% and a specificity of 71% were found for a cutoff point of 105 CFU/mL in the ETA, showing good agreement with the BAL results. SHIN et al. (2011) obtained even better results, with 85.7% sensitivity and 94.7% specificity for ETA culture for a cutoff point of 106 CFU/mL, a higher yield than that obtained with BAL culture.

The study by DiGiovine et al. (1999) aimed to assess attributable mortality and costs associated with an episode of BSI. These authors found no differences in relation to a control group of non-infected patients regarding an association between BSI and mortality, but there were significant differences with costs and length of stay in the ICU.

Regarding polytraumatized patients, Glance et al. (2011) found that length of stay and costs were significantly higher for those patients who had infections within the hospital. According to these authors, patients with nosocomial infections may have a longer hospital stay, and prolonged hospitalization may increase the risk of acquiring HAI. They also make an inference regarding the association between HAIs and cost, as length of stay is an important component of cost.

Two studies focused on length of stay. In the first study, by Barnett et al. (2013), increased length of stay is seen because of BSI. This work was carried out with patients not exclusively admitted to the ICU, but provided some interesting data, such as: the average number of extra days of hospitalization was lower in the ICU than in the wards; BSI due to coagulase-negative staphylococci were responsible for the longest extra period of ICU stay; extra days of stay were generally longer for patients who died. In the second study, by Jeon et al. (2012), increased length of stay may be associated with BSI. This association was demonstrated by the presence of underlying diseases and the need for invasive procedures, increasing hospitalization time and the risk of developing BSI. The approaches of these two studies show that there is a reverse association between risk of HI and increased length of stay, that is, it is a two-way street, in the words of Angelis et. al. (2010) and Kollef et al. (2012). Another infection also related to a prolonged hospital stay is CDI, which results from alteration of the normal intestinal microbiota due to the use of broad-spectrum antimicrobials. O'Brien et al. (2007) report an average increase in hospitalization of 2.9 to 3.6 days, requiring additional antimicrobial treatment for ten to 14 days.

Multidrug resistant (MDR) microorganisms are independent factors in increasing length of stay and hospital costs. These increases may be due to a delay in treatment with effective antimicrobials (COSGROVE, 2006; SWANSON; WELLS, 2013) or also due to the need to perform surgeries and other procedures (COSGROVE, 2006).

In agreement with the results of Cosgrove (2006), three patients in the present series presented bacteremia due to *Staphylococcus aureus.* The two strains sensitive to Oxacillin (MSSA) were from infections prior to ICU admission, and the third strain, resistant to Oxacillin (MRSA), was from an HAI acquired in the ICU. The two patients with sensitive strains remained in the ICU for four and seven days, while the person with the MDR strain was hospitalized for 39 days. Each microorganism has its own multidrug resistance pattern, as shown in figure 11.

Lodise et. al. (2003, apud COSGROVE, 2006) found that a longer time to effective treatment of nosocomial bacteremia caused by *Staphylococcus aureus* was an independent predictor of infection-related mortality and was associated with a longer duration of hospitalization after resolution of the bacteremia.

The average length of stay in the adult ICU is an extremely important indicator for the functioning of this sector of the hospital, as it measures the efficiency of bed management. The lack of beds in this sector can constitute a bottleneck for the admission of patients from hospitalization units, the surgical center and the emergency room. The average length of stay in the country's ICUs varies from three to four days, according to Silva (2007 apud Agência Nacional de Saúde Suplementar[6], 2013). In 58.1% of 47 ICUs in the city of São Paulo, this indicator was 4.5 days (KIMURA; KOIZUMI; MARTINS, 1997 apud Agência Nacional de Saúde Suplementar, 2013).

Score-based systems have been used for a long time with the aim of providing prognoses about

---

[6] Agência Nacional de Saúde Suplementar is a special authority with administrative autonomy, responsible for supervising health plan operators and regulating the market, both in terms of assistance and those linked to economic activity.

clinical outcome in a more objective way. The APACHE II scale, in use since 1985 and well known by intensivists, is one of these systems. According to Suka, Yoshida and Takezawa (2004), the scale is based on twelve routine physiological measurements, age and previous health status (presence or not of chronic organic failure, presence or not of elective or emergency surgery). The physiological measurements evaluated by this scale are body temperature, respiratory and heart rates, mean arterial pressure, the Glascow scale, the presence or absence of acute renal failure, serum creatinine, hematocrit, leukometry, and dosages. serum levels of sodium, potassium and bicarbonate. If arterial blood gas analysis is available, arterial pH and arterial O2 pressure (PaO2) are also included in the calculation to determine the index and, consequently, mortality. When the patient is sedated, it is not possible to determine the Glascow scale.

When comparing the results found among the 54 patients in this series who had APACHE II calculated, with the results of Chiavone and Sens (2003), it can be observed that there was almost agreement regarding progressively increasing mortality (figure 18). The differences occurred in the 16-20 and 26-30 score ranges of the present case series, which showed higher mortality rates than in the 21-25 and above 30 ranges, respectively. These differences may have occurred due to the small number of patients in the present sample.

In agreement with the results of Suka, Yoshida and Takezawa (2004), the number of HAIs acquired in the ICU increased by 133%, directly, between categories 0-16 and 17-41 of the APACHE II score. It can be said that the APACHE II scale is a good predictor of nosocomial infections in the ICU.

As demonstrated by Graves (2004), there is no proportional correspondence between the amount invested in HAI prevention and the results of a prevention program. Considering from an economic and administrative perspective, there is an "optimal" point of incidence of HAI that minimizes the total variable costs of hospitalizations as much as possible, making it compatible with the amounts invested, which would be the ideal objective of every policymaker.

HAIs acquired during the ICU stay do not constitute, in themselves, a cause of increased length of stay and, consequently, costs. Invasive mechanical ventilation (IMV) is also associated with a prolonged stay, with around ⅓ of patients admitted to an ICU requiring this modality of ventilatory support (ESTEBAN et. al., 2002; HEBERT et al., 2001 apud ROTTA et al., 2018). Of the 76 patients in the present series from whom it was possible to obtain information about IMV, 43 patients had used it and remained hospitalized for an average of 12.4 days, while 33 patients who did not use IMV were hospitalized for an average of 7.2 days.

Invasive mechanical ventilation is associated with higher costs, representing up to 12% of a hospital's total expenditure (ROTTA et. al., 2018). These authors emphasize the need for a physiotherapist 24 hours a day in the ICU, which can contribute to reducing the time required to use invasive mechanical ventilation, reducing treatment costs.

An intervention aimed at reducing the incidence of HAIs in the ICU and hospital costs is oral hygiene with chlorhexidine solution (OHC). Chlorhexidine is effective in reducing dental plaque, reducing the concentration of microorganisms in the oral cavity, having an important effect in preventing VAP (ATAY; KARABACAK, 2014). OHC reduces the risk of developing VAP from 24% to 18%, that is, one case of VAP can be prevented in every 17 patients with invasive mechanical ventilation (HUA et. al., 2016).

Another measure that can contribute to reducing the length of the hospital stay and the resulting costs is the speed in identifying and communicating the infection to the care staff. Beekmann et al. (2003) carried out a study with patients presenting BSI and found that the use of automated systems for continuous monitoring of blood cultures shortened the time needed to obtain a positive result, impacting the length of stay in hospital.

Antimicrobial control programs (Antimicrobial Stewardship Program or ASP) also enable positive results in hospital costs. The Centre Hospitalier Universitaire de Sherbrooke, a Canadian university hospital, adopted a computerized clinical decision support system. At the end of five years (August 2008 to August 2013) between implementation of the program and analysis of results, there was the adoption of best practices by the clinical staff, favoring the optimization of antimicrobial treatment and resulting in shorter periods of hospitalization and shorter costs (NAULT et al., 2017).

The concept of Antimicrobial Stewardship Program refers to a strategy that aims to improve the use of antimicrobials with the objectives of improving clinical outcomes, reducing antimicrobial costs and minimizing adverse effects associated with the use of these medicines, including microbial resistance. In the systematic review carried out by Karanika et al. (2016) the use of ASP allowed a general reduction in antimicrobial consumption of 20%, this effect being more pronounced in the ICU, where there was a decrease of around 40%. Furthermore, there was a reduction in costs with the use of antimicrobials, length of stay in the ICU, and infections with MRSA, imipenem-resistant *Pseudomonas aeruginosa* and *Klebsiella spp.* producers of ESBL (extended spectrum beta-lactamases).

According to Jayaram and Ramakrishnan (2008), the Emergency maintains a close collaborative relationship with the ICU as it is the gateway to the hospital for many patients. By adequately stabilizing the most serious cases, the Emergency provides greater possibilities for higher quality subsequent care, a favorable clinical outcome, and reduced costs.

CONCLUSION

This monograph aimed to show the economic repercussions of care-related infections acquired during patients' stay in an intensive care unit of a medium-sized tertiary public hospital. To this end, the methodology chosen was a comparison between patients who acquired HAIs in the ICU, those who did not develop an infection, and those who already had an infectious condition upon admission to this sector.

As shown in the results, the average number of days of hospitalization in the group of 12 patients with HAIs acquired in the ICU was 3.1 times higher compared to the group of 33 patients who had already been admitted with infection, and 4.1 times higher compared to the group of 33 patients who did not present infection. This prolonged average length of stay was accompanied by an increase in hospitalization costs, where 15.4% of patients (the twelve with HAIs acquired in the ICU) corresponded to 38.4% of the total costs with the 78 patients in the study.

The presence of an infection with a multidrug-resistant germ also correlated with an increased length of stay, with an example being presented as a comparison between an infection by *S. aureus* MRSA and two others by MSSA.

However, to understand the factors that lead to the acquisition of HAIs in an ICU environment, it was necessary to show the variety of microorganisms involved, the frequency of invasive procedures performed during the period under study and the influence of APACHE II scale severity scores on outcomes. clinicians. It was also necessary to present the criteria for diagnosing the main infections (associated with mechanical ventilation, the use of indwelling bladder catheters, and the use of intravascular devices) and the instrument (epidemiological surveillance form) through which information was collected along with to patients.

Literature on the subject has been growing steadily in several countries in Europe, North America and Asia. However, there are still few national contributions. This was one of the motivations that led the author of the monograph, a bachelor in Hospital Administration, to choose this theme.

There were limitations to the study. Firstly, the study was retrospective, based on the analysis of primary sources of information. This methodology has the advantages of easy access to original data, which have not yet been consolidated, and the author can then construct the information in his own way. The disadvantages are the lack of monitoring of variables and sometimes the data is inconsistent or not reported. The second limitation, imposed by the short time available for data collection, was the small sample size on which the study was carried out. Although the objectives outlined before the start of the research are considered achieved, it is known that the larger the sample, the greater the accuracy of the information extracted. An example of the lack of accuracy in this study was the non-perfectly progressive increase in the risk of mortality according to the APACHE II scale.

Other limitations, now in relation to the studies researched that served as theoretical references, were the great variability of methodologies used, as well as the scarcity of common denominators and correction factors that did not allow a comparison of the findings of this study with those of other countries and through the time.

On the other hand, the study undertaken presents some possibilities and opportunities. The main possibility was to highlight the magnitude and relevance that infections related to intensive care can assume in relation to treatment costs, which can help health managers to plan and efficiently use resources with a view to offer high quality care. An opportunity, considered important by the author of this study, would be to carry out more in-depth studies on the various variables related to HAIs that influence the costs of intensive care, whether directly or indirectly.

BIBLIOGRAPHY

ALP E, KALIN G, COSKUN R, et al. Economic burden of ventilator-associated pneumonia in a developing country. J Hosp Infect, v. 81, n. 2, p. 128-30, 2012.

ANGELIS G, MURTHY A, BEYERSMANN J, et al. Estimating the impact of healthcare-associated infections on length of stay and costs. Clin Microbiol Infect, v. 16, n. 12, p. 1729-35, 2010.

ATAY S, KARABACAK Ü. Oral care in patients on mechanical ventilation in intensive care unit: literature review. Int J Res Med Sci, v. 2, n. 3, p. 822-9, 2014.

BARNETT AG, PAGE K, CAMPBELL M, et al. The increased risks of death and extra lengths of hospital and ICU stay from hospital-acquired bloodstream infections: a case-control study. BMJ Open, v. 3, n. 10, p. 1-6, 2013.

BEEKMANN SE, HENDERSON DK. Infections caused by percutaneous intravascular devices. In: MANDELL, GL; BENNETT, JE; DOLIN, R (Org.). Principles and practice of infectious diseases. 6th ed. Philadelphia: ELSEVIER Churchill Livingstone, 2005. v. 2, chapter 300, p. 3347-62.

BERNARD H, HACKBARTH D, OLMSTED RN, et al. Creation of a competency-based professional development program for infection preventionists guided by the APIC Competency Model: steps in the process. Am J of Infect Control, v. 46, n. 11, p. 1202-10, 2018.

BILLINGS C, BERNARD H, CAFFERY L, et al. Advancing the profession: an updated future-oriented competency model for professional development in infection prevention and control. Am J of Infect Control, v. 47, n. 6, p. 602-14, 2019.

BRAGA, MA. Influence of care-related infections on length of stay and hospital mortality using the diagnosis related groups classification as clinical risk adjustment. Belo Horizonte: Faculty of Medicine of the Federal University of Minas Gerais, 2015. 132 p. Doctorate Thesis, UFMG, Belo Horizonte, 2015.

BRAZIL. Ministry of Health. Ordinance GM/MS nº 2.616, May 12, 1998. Guidelines and standards for the prevention and control of hospital infections. Official Gazette [of] the Federative Republic of Brazil, Brasília, n. 89, p. 133-5, May 13, 1998. Section 1.

BRAZIL. Ministry of Health. National Supplementary Health Agency. Average adult ICU stay. Brasília, v. 1.01, January 2013.

BRAZIL. National Health Surveillance Agency. Diagnostic Criteria for Healthcare-Associated Infections. 2nd ed. Brasília: ANVISA, 2017. 135 p.

BRAZILIAN INTENSIVE MEDICINE ASSOCIATION. AMIB Census 2016. Brazilian ICU.

CARVALHO MVCF, WINKELER GFP, COSTA FAM, et al. Agreement between tracheal aspirate and bronchoalveolar lavage in the diagnosis of pneumonia associated with mechanical ventilation. J Bras Pneumol, v. 30, n. 1, p. 26-38, 2004.

CHIAVONE PA, SENS YAS. Evaluation of APACHE II system among intensive care patients at a teaching hospital. São Paulo Med J, v. 121, n. 2, p. 53-7, 2003.

CORRÊA RA, LUNA CM, ANGELS JCFV, et. al. Quantitative culture of tracheal aspirate and bronchoalveolar lavage in the management of patients with ventilator-associated pneumonia: a randomized clinical trial. J Bras Pneumol, v. 40, n. 6, p. 643-51, 2014.

COSGROVE SE. The relationship between antimicrobial resistance and patient outcomes: mortality, length of hospital stay, and health care costs. Clin Infect Dis, v. 42, n. Supplement_2, p. S82-S89, 2006.

CHACKO B, THOMAS K, DAVID T, et al. Attributable cost of a nosocomial infection in the intensive care unit: a prospective cohort study. World J Crit Care Med, v. 6, n. 1, p. 79-84, 2017.

CZAJKA R, SALYER C. Margin of excellence: intensive care unit (ICU) utilization. Premier Consulting, Charlotte, 2017.

DAVIS J, BILLINGS C, MALIK C. Revisiting the Association for Professionals in Infection Control and Epidemiology Competency Model for the Infection Preventionist: an evolving conceptual framework. Am J of Infect Control, v. 46, n. 8, p. 921-7, 2018.

DiGIOVINE B, CHENOWETH C, WATTS C, et al. The attributable mortality and costs of primary nosocomial bloodstream infections in the intensive care unit. Am J Respir Crit Care Med, v. 160, n. 3, p. 976-81, 1999.

ERBAY RH, YALCIN AN, ZENCIR M, et al. Costs and risk factors for ventilator-associated pneumonia in a Turkish university hospital's intensive care unit: a case-control study. BMC Pulmonary Medicine, v. 4, n. 3, p. 1-7, 2004.

FEDERAL COUNCIL OF MEDICINE. Less than 10% of Brazilian municipalities have an ICU bed through the SUS, according to a CFM survey.

GLANCE LG, STONE PW, MUKAMEL DB, et al. Increases in mortality, length of stay, and costs associated with hospital-acquired infections in trauma patients. Arch Surg, v. 146, n. 7, p. 794-801, 2011.

GLIED S, COHEN B, LIU J, et al. Trends in mortality, length of stay and hospital charges associated with healthcare-associated infections, 2006-2012. Am J Infect Control, v. 4, n. 9, p. 983-9, 2016.

GRAVES N. Economics and preventing hospital-acquired infection. Emerg Infect Dis, v. 10, n. 4, p. 561-6, 2004.

HENMAN LJ, CORRIGAN R, CARRICO R, et al. Identifying changes in the role of the infection preventionist through the 2014 practice analysis study conducted by the Certification Board of Infection Control and Epidemiology, Inc. Am J of Infect Control, v. 43, n. 7, p. 664-8, 2015.

HUA F, XIE H, WORTHINGTON HV, et al. Oral hygiene care for critically ill patients to prevent ventilator-associated pneumonia (intervention review) Cochrane Database of Systematic Rev, n. 10, 2016.

JAIN M, MILLER L, BELT D, et al. Decline in ICU adverse events, nosocomial infections and costs through a quality improvement initiative focusing on teamwork and culture change. Qual Saf Health Care, v. 15, n. 4, p. 235-9, 2006.

JAYARAM R, RAMAKRISHNAN N. Cost of intensive care in India. Indian J Crit Care Med, v. 12, n. 2, p. 55-61, 2008.

JEON CY, MATTHEW N, HAOMIAO J, et al. On the role of length of stay in healthcare-associated bloodstream infection. Infect Control Hosp Epidemiol, v. 33, n. 12, p. 1213-8, 2012.

KARANIKA S, PAUDEL S, GRIGORAS C, et al. Systematic review and meta-analysis of clinical and economic outcomes from the implementation of hospital-based antimicrobial stewardship programs. Antimicrob Agents Chemother, v. 60, n. 8, p. 4840-52, 2016.

KERLIN MP, COOKE CR. Understanding costs when seeking value in critical care. Ann Am Thorac Soc, v. 12, n. 12, p. 1733-4, 2015.

KOLLEF MH, HAMILTON CW, ERNST FR. Economic impact of ventilator-associated pneumonia in a large matched cohort. Infect Control Hosp Epidemiol, v. 33, n. 3, p. 250-6, 2012.

LIMA ME, ANDRADE D, HAAS VJ. Prospective evaluation of the occurrence of infection in critically ill patients in the Intensive Care Unit. Rev Bras Ter Intensiva, v. 19, n. 3, p. 342-7, 2007.

MARTINS MA. General historical aspects. In: _____. Hospital infection manual: epidemiology, prevention and control. 2nd ed. Rio de Janeiro: MEDSI, 2001a. Chapter 1, p. 3-10.

MARTINS MA. General concepts and basic terminology in hospital epidemiology and infection control. In: _____. Hospital infection manual: epidemiology, prevention and control. 2nd ed. Rio de Janeiro: MEDSI, 2001b. Chapter 3, p. 16-26.

MATHAI AS, PHILLIPS A, KAUR P, et al. Incidence and attributable costs of ventilator-associated pneumonia (VAP) in a tertiary-level intensive care unit (ICU) in northern India. J Infect Public Health, v. 8, n. 2, p. 127-35, 2015.

McCAULEY LM, WEBB BJ, SORENSEN J, et al. Use of tracheal aspirate culture in newly intubated patients with community-onset pneumonia. Ann Am Thorac Soc, v. 13, n. 3, p. 376-81, 2016.

NANGINO GO, OLIVEIRA CD, CORREIA PC, et al. Financial impact of nosocomial infections in intensive care units in a philanthropic hospital in Minas Gerais. Rev Bras Ter Intensiva, v. 24, n. 4, p. 357-61, 2012.

NAULT V, PEPIN J, BEAUDOIN M, et al. Sustained impact of a computer-assisted antimicrobial stewardship intervention on antimicrobial use and length of stay. J Antimicrobiol Chemother, v. 72, n. 3, p. 933-40, 2017.

NIEDERMAN, MS. Cost effectiveness in treating ventilator-associated pneumonia. Critical Care, v. 5, n. 5, p. 243-4, 2001.

O'BRIEN JA, LAHUE BJ, CARO JJ, et al. The emerging infectious challenge of Clostridium difficile-associated disease in Massachusetts hospitals: clinical and economic consequences. Infect Control Hosp Epidemiol, v. 28, n. 11, p. 1219-27, 2007.

OLAECHEA PM, ULIBARRENA MA, ALVAREZ-LERMA F, et al. Factors related to hospital stay among patients with nosocomial infection acquired in the intensive care unit. Infect Control Hosp Epidemiol, v. 24, n. 3, p. 207-13, 2003.

OLIVEIRA ABF, DIAS OM, MELLO MM, et al. Factors associated with higher mortality and prolonged length of stay in an adult intensive care unit. Rev Bras Ter Intensiva, v. 22, n. 3, p. 250-6, 2010.

PADOVEZE MC; FORTALEZA CMCB. Healthcare-related infections: challenges for public health in Brazil. Rev Saúde Pública, v. 48, n. 6, p. 995-1001, 2014.

PRADHAN NP, BHAT SM, GHADAGE DP. Nosocomial infections in the medical ICU: a retrospective study highlighting their prevalence, microbiological profile and impact on ICU stay and mortality. J Assoc Physicians India, v. 62, n. 10, p. 18-21, 2014.

PUGLIESE G. Public Health focus: surveillance, prevention, and control of nosocomial infections. MMWR, v. 41, n. 42, p. 783-7, 1992.

ROTTA BP, SILVA JM, FU C, et al. Relationship between availability of physiotherapy services and ICU costs. J Bras Pneumol, v. 44, n. 3, p. 184-9, 2018.

SAFDAR N, DEZFULIAN C, COLLARD HR, et al. Clinical and economic consequences of ventilator-associated pneumonia: a systematic review. Crit Care Med, v. 33, n. 10, p. 2184-93, 2005.

SCOTT II RD. The direct medical costs of healthcare-associated infections in U.S. hospitals and the benefits of prevention. Division of Healthcare Quality Promotion, National Center for Preparedness, Detection, and Control of Infectious Diseases, Coordinating Center for Infectious Diseases, Centers for Disease Control and Prevention. 2009.

SHIN YM, YEON-MOK O, KIM MN, et al. Usefulness of quantitative endotracheal aspirate cultures in intensive care unit patients with suspected pneumonia. J Korean Med Sci, v. 27, n. 7, p. 865-9, 2011.

STONE PW. Economic burden of healthcare-associated infections: an American perspective. Expert Rev Pharmacoecon Outcomes Res, v. 9, n. 5, p. 417-22, 2009.

STRAUSBAUGH LJ. Nosocomial respiratory infections. In: MANDELL GL, BENNETT JE, DOLIN R. (Org.). Principles and practice of infectious diseases. 6th ed. Philadelphia: ELSEVIER Churchill Livingstone, 2005. v. 2, chapter 301, p. 3362-70.

SWANSON JM, WELLS DL. Empirical antibiotic therapy for ventilator-associated pneumonia. Antibiotics, v. 2, n. 3, p. 339-51, 2013.

SUKA M, YOSHIDA K, TAKEZAWA J. Association between APACHE II score and nosocomial infections in intensive care unit patients: a multicenter cohort study. Environ Health Prev Med, v. 9, n. 6, p. 262-5, 2004.

TAN SS, BAKKER J, HOOGENDOORN ME, et al. Direct cost analysis of intensive care unit stays in four European countries: applying a standardized costing methodology. Value in Health, v. 15, n. 1, p. 81-6, 2012.

THOMAS K, PETER JV, CHRISTINA J, et al. Cost utility in medical intensive care patients: rationalizing ongoing care and timing of discharge from intensive care. Ann Am Thorac Soc, v. 12, n. 7, p. 1058-65, 2015.

TIWARI P, ROHIT M. Assessment of costs associated with hospital-acquired infections in a private tertiary care hospital in India. Value Health Reg Issues, v. 2, p. 87-91, 2013.

WARREN JW. Nosocomial urinary tract infections. In: MANDELL GL, BENNETT JE, DOLIN R. (Org.). Principles and practice of infectious diseases. 6th ed. Philadelphia: ELSEVIER Churchill Livingstone, 2005. v. 2, chapter 302, p. 3370-81.

ZIMLICHMAN E, HENDERSON D, TAMIR O, et al. Health care-associated infections: a meta-analysis of costs and financial impact on the US health care system. JAMA Intern Med, v. 173, n. 22, p. 2039-46, 2013.

APPENDICES

Figure 03 – Patient transit within the hospital

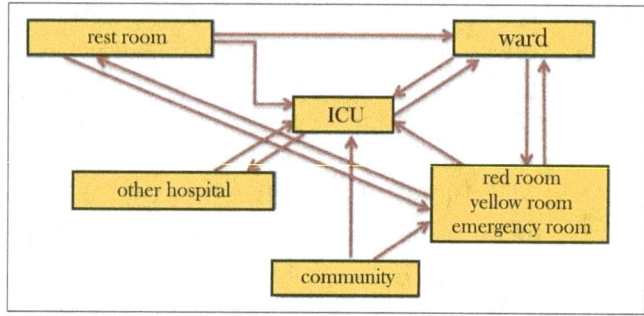

Figure 04 – Distribution of patients according to the presence or absence of infection

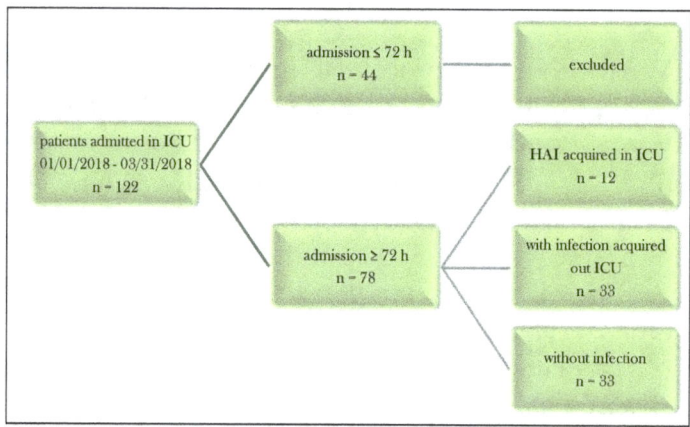

Figure 05 – Days spent in the Intensive Care Unit

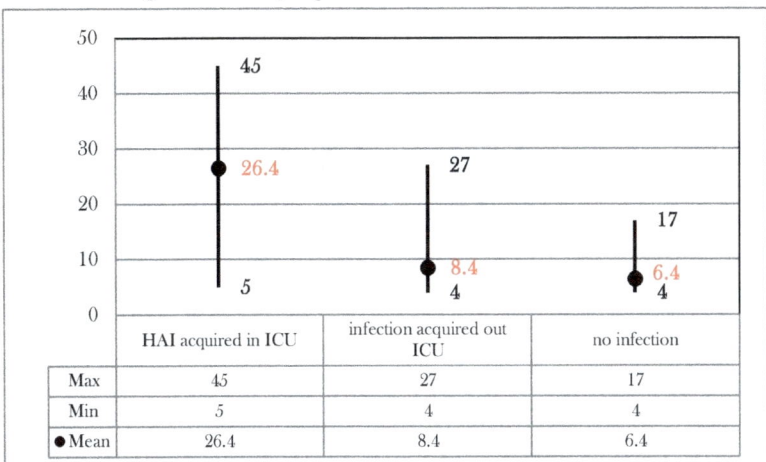

	HAI acquired in ICU	infection acquired out ICU	no infection
Max	45	27	17
Min	5	4	4
● Mean	26.4	8.4	6.4

Fig 06 – Days of stay until discharge from the Intensive Care Unit

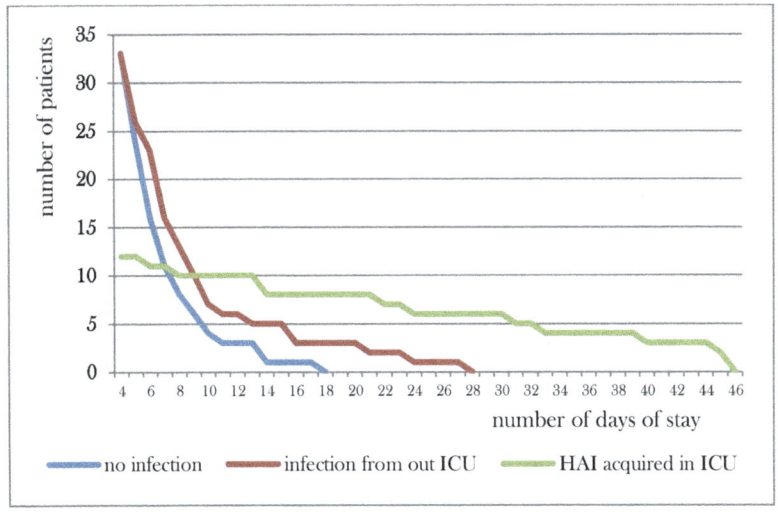

Figure 07 – Frequencies of all invasive procedures

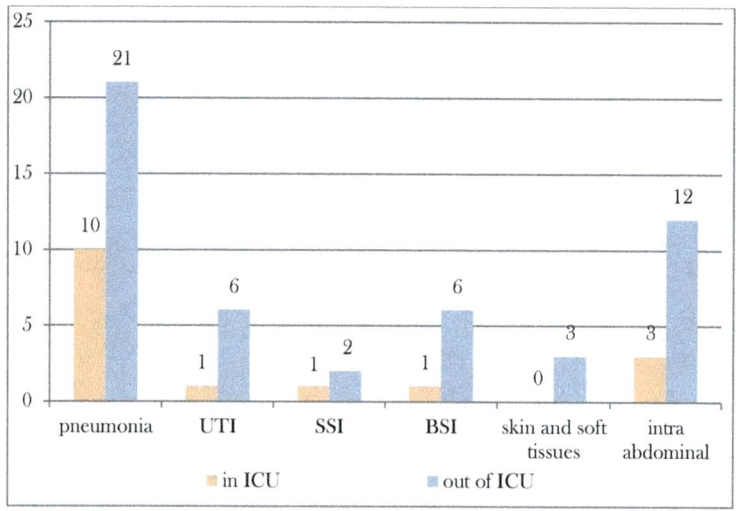

	urinary catheter	peripheral access	central catheter	OTT/tracheostomia	gastric catheter	enteral catheter	dialysis catheter	PEN	MAP
no	18	33	34	35	43	54	72	73	77
yes	60	45	44	43	35	24	6	5	1

Figure 08 – Topographies of infectious processes

(bar chart)

pneumonia — in ICU: 10, out of ICU: 21
UTI — in ICU: 1, out of ICU: 6
SSI — in ICU: 1, out of ICU: 2
BSI — in ICU: 1, out of ICU: 6
skin and soft tissues — in ICU: 0, out of ICU: 3
intra abdominal — in ICU: 3, out of ICU: 12

■ in ICU ■ out of ICU

Fig 09 – Distribution of microorganisms (ICU-acquired HAIs)

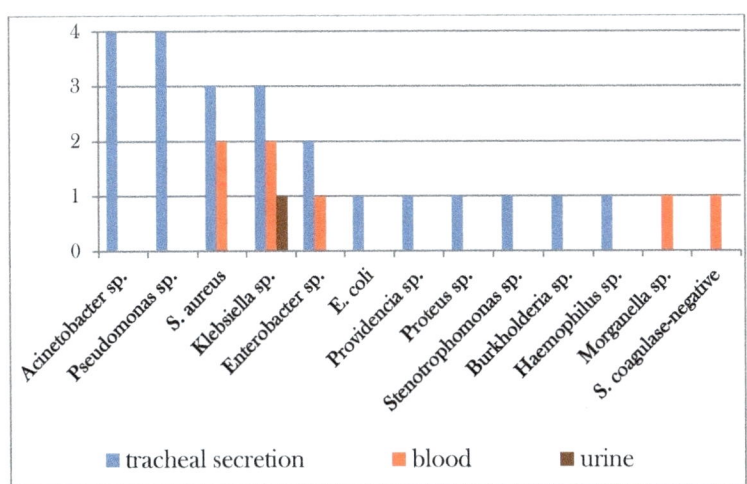

Fig 10 – Distribution of microorganisms (infections present upon admission to the ICU)

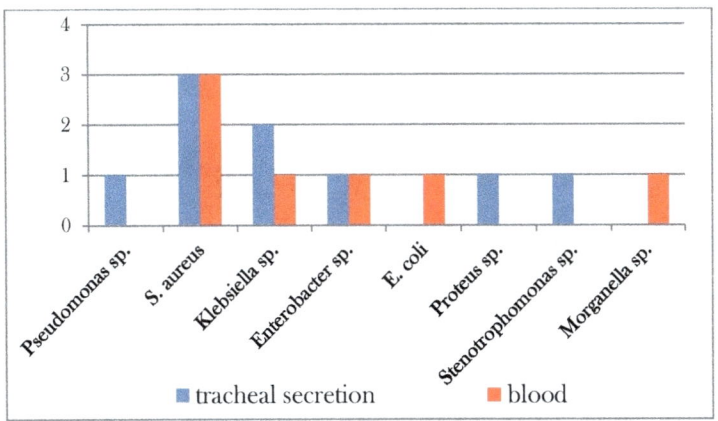

Figure 11 – Multidrug Resistance

Pseudomonas aeruginosa	R to carbapenem R to amikacin R to 3rd and 4th generations cephalosporins
Enterococcus sp.	R/I to vancomycin or teicoplanin
Acinetobacter sp.	R to carbapenem R to ampicillin/sulbactam
Enterobacter sp.	ESBL producers R to 3rd and 4th generations cephalosporins R to aztreonam R to carbapenem R to aminoglycosides and 3rd generation cephalosporins
Staphylococcus aureus	R to oxacillin R/I to vancomycin or teicoplanin
Coagulase-negative *Staphylococcus*	R/I to vancomycin or teicoplanin
Burkholderia cepacia	All identified
Clostridioides difficile	All identified
Stenotrophomonas maltophila	R to trimethoprim-sulfamethoxazole

Obs: R (resistance) / I (intermediate resistance)

Fig 12 – Average expenditure (R$) of each group with each component of the costs[7]

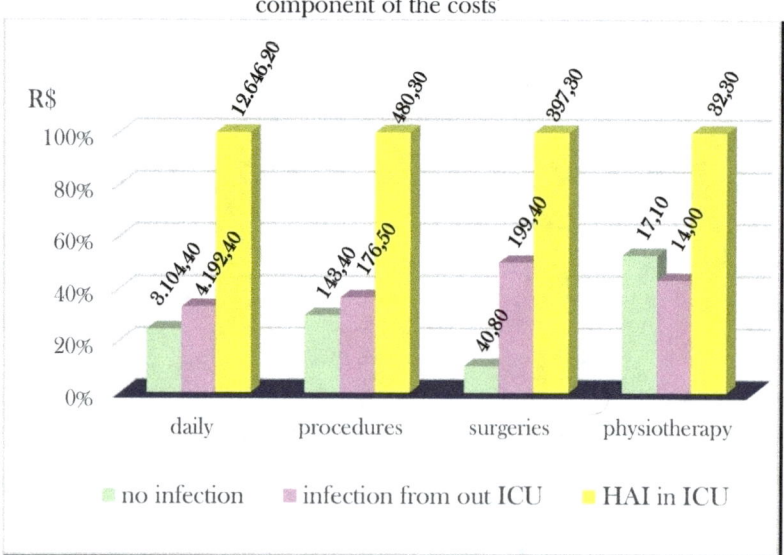

[7] Numerical representation of monetary values in Brazilian Reais (R$).

ATTACHMENTS

Fig 13 – Defining criteria for Surgical Site Infection

	Superficial incisional SSI	Deep incisional SSI
Appearance time	It occurs in the first 30 days after the surgical procedure (day 1 being the date of the procedure)	It occurs in the first 30 days after surgery (the 1st day being the date of the procedure) or up to 90 days if implants are placed
Involvement	involves only the skin and subcutaneous tissue	involves soft tissues deep to the incision (e.g. fascia and/or muscles)
Criteria	at least ONE of the following criteria: • Purulent drainage from the superficial incision; • Positive culture of secretion or tissue from the incision, obtained aseptically; • The superficial incision is deliberately opened by the surgeon in the presence of at least one of the following signs or symptoms: pain, increased sensitivity, local edema, hyperemia or heat, unless the culture is negative; • Diagnosis of infection by the surgeon or other attending physician.	at least ONE of the following criteria: • Purulent drainage from the deep incision, but not originating from an organ/cavity; • Deep spontaneous dehiscence or incision opened by the surgeon and culture positive or not performed, when the patient presents at least 1 of the following signs and symptoms: fever, pain or local swelling • Abscess or other evidence of infection involving deep tissues, detected during clinical, anatomopathological or imaging examination; • Diagnosis of deep incisional infection made by the surgeon or other attending physician
Types	• Primary superficial incisional: identified in the primary incision in a patient with more than 1 incision • Secondary superficial incisional: identified in the secondary incision in a patient with more than 1 incision	• Primary deep incisional: identified in the primary incision in a patient with more than 1 incision • Secondary deep incisional: identified in the secondary incision in patients with more than 1 incision
Comments	In the case of ophthalmic surgery, conjunctivitis will be defined as a superficial incisional infection; Do not report minimal inflammation and drainage of secretion limited to the suture points; The following are not defined as superficial SSI: • Diagnosis or treatment of cellulitis. An incision that is drained or has a microorganism identified in culture or by a molecular diagnostic method is not considered cellulitis; • Surgical point abscess • Infection from episiotomy or circumcision of the newborn.	
	Source: National Health Surveillance Agency, Diagnostic Criteria for Healthcare-Associated Infections	

Adapted from: **BRASIL**. Agência Nacional de Vigilância Sanitária. *Diagnostic Criteria for Healthcare-Associated Infections*)

Fig 14 – Flowchart for reporting BSI associated with central catheters

1. Patient has a recognized pathogen (not a common skin contaminant) cultured from one or more blood cultures AND the microorganism is not related to an infection elsewhere?					
YES	2. Does the patient have a central catheter on the date of infection OR was it removed up to one day before the date of infection AND was the central catheter inserted for more than 2 days on the date of infection?				
	YES	Fulfills criterion 1 for IPCSL associated with central catheter			
	NO	Does not meet criteria for IPCSL associated with central catheter			
NO	2. Patients with two or more blood cultures (in different punctures with a maximum interval of 2 days) with a common skin contaminant AND the microorganism is not related to an infection elsewhere?				
	NO	Does not meet criteria for IPCSL associated with central catheter			
	YES	3.1. Patient > 28 days and ≤ 1 year?			
		YES	4. At least one of the following signs and symptoms (fever, hypothermia, apnea, bradycardia) 3 days before or 3 days after the positive blood culture?		
			YES	5. Does the patient have a central catheter on the date of infection OR was it removed up to 1 day before the date of infection AND was the central catheter inserted > 2 days on the date of infection?	
				YES	Fulfills criterion 3 for IPCSL associated with central catheter
				NO	Does not meet criteria for IPCSL associated with central catheter
			NO	Does not meet criteria for IPCSL associated with central catheter	
		NO	Does not meet criteria for IPCSL associated with central catheter		
	YES	3.2. Patient > 1 year old?			
		YES	4. At least one of the following signs and symptoms (fever, chills, arterial hypotension) 3 days before or 3 days after the positive blood culture?		
			YES	5. Does the patient have a central catheter on the date of infection OR was it removed up to 1 day before the date of infection AND was the central catheter inserted > 2 days on the date of infection?	
				YES	Fulfills criterion 2 for IPCSL associated with central catheter
				NO	Does not meet criteria for IPCSL associated with central catheter
			NO	Does not meet criteria for IPCSL associated with central catheter	
		NO	Does not meet criteria for IPCSL associated with central catheter		

Adapted from: BRASIL. Agência Nacional de Vigilância Sanitária. *Diagnostic Criteria for Healthcare-Associated Infections*)

Fig 15A – Diagnostic criteria for pneumonia associated with mechanical ventilation

Fig 13 – Defining criteria for Pneumonia associated to Mechanical Ventilation

	Clinically defined pneumonia	Microbiologically defined pneumonia
Patient with underlying cardiac or pulmonary disease with ≥ 2 serial chest radiographs with one of the following findings, persistent, new or progressive: • Infiltrate; • Opacification; • Cavitation And at least ONE of the following signs and symptoms: • Fever temperature > 38°C, with no other associated cause; • Leukopenia (<4,000 cells/mm³) or leukocytosis (> 12,000 cells/mm³); • Change in the level of consciousness, with no other apparent cause, in patients ≥ 70 years old	✓	✓
And at least TWO of the following signs and symptoms: • Appearance of purulent secretion or change in the characteristics of the secretion or increase in respiratory secretion or increased need for aspiration; • Worsening gas exchange (desaturation, such as PaO_2/FiO_2 < 240) or increased oxygen supply or increased ventilatory parameters); • Auscultate with rhonchi or rales; • Onset or worsening of cough or dyspnea or tachypnea.	✓	
And at least ONE of the following signs and symptoms: • Appearance of purulent secretion or change in the characteristics of the secretion or increase in respiratory secretion or increased need for aspiration; • Worsening gas exchange (desaturation, such as PaO_2/FiO_2 < 240) or increased oxygen supply or increased ventilatory parameters); • Auscultate with rhonchi or rales; Onset or worsening of cough or dyspnea or tachypnea.		✓
At least ONE of the results below: • Positive blood culture, without another source of infection; • Positive culture of pleural fluid; • Positive quantitative culture of the secretion obtained by a procedure with less potential for contamination (e.g. BAL or protected brushing); • Finding of ≥ 5% of leukocytes and macrophages containing microorganisms (intracellular bacteria on BAL bacterioscopy); • Positive lung tissue culture		

Fig 15B – Diagnostic criteria for pneumonia associated with mechanical ventilation

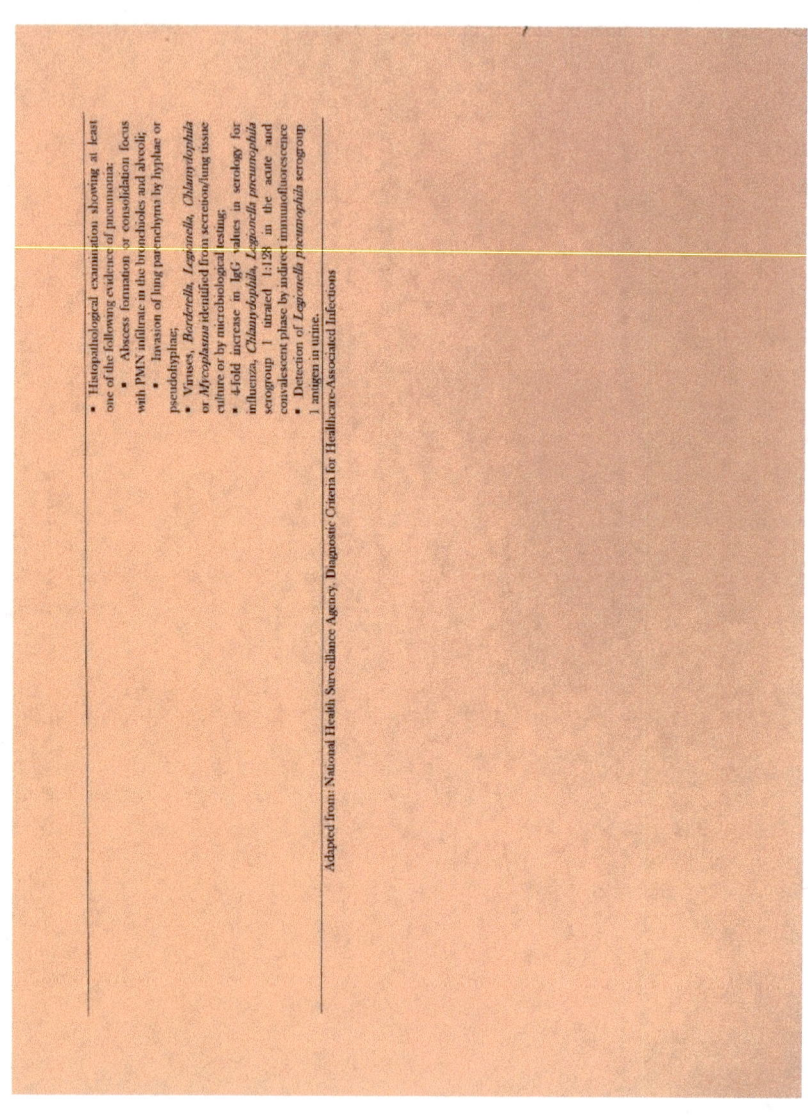

- Histopathological examination showing at least one of the following evidence of pneumonia:
  - Abscess formation or consolidation focus with PMN infiltrate in the bronchioles and alveoli;
  - Invasion of lung parenchyma by hyphae or pseudohyphae;
- *Viruses, Bordetella, Legionella, Chlamydophila* or *Mycoplasma* identified from secretion/lung tissue culture or by microbiological testing;
- 4-fold increase in IgG values in serology for influenza, *Chlamydophila, Legionella pneumophila* serogroup 1 titrated 1:128 in the acute and convalescent phase by indirect immunofluorescence
- Detection of *Legionella pneumophila* serogroup 1 antigen in urine.

Adapted from: National Health Surveillance Agency. Diagnostic Criteria for Healthcare-Associated Infections

Fig 16 - Epidemiological Surveillance Bulletin (front side)

EPIDEMIOLOGICAL SURVEILLANCE BULLETIN

Month:		Bed/Sector:	Time:
Name:	Age:	Date of hospital admission:	Entry date into the sector:
Date of leaving the sector:	Hour:	internal transfer ( ) external transfer ( ) death ( )	Related to hospital infection: ( ) Y ( ) N

Admission diagnosis:		Sector origin:
Final diagnosis:	Medical record:	Surgery date:

Date
Temperature
Leukocytes

Invasive Procedures:

Clinical signs:

( )	( )	( )	( )
( )	( )	( )	( )
( )	( )	( )	( )
( )	( )	( )	( )
( )	( )	( )	( )
( )	( )	( )	( )
( )	( )	( )	( )
( )	( )	( )	( )

Topography of infection:

( ) Pneumonia	( )
( ) Urinary tract infection	( )
( ) Surgical site infection	( ) Other (which?)
( )	

Fig 17 – Epidemiological Surveillance Bulletin (back side)

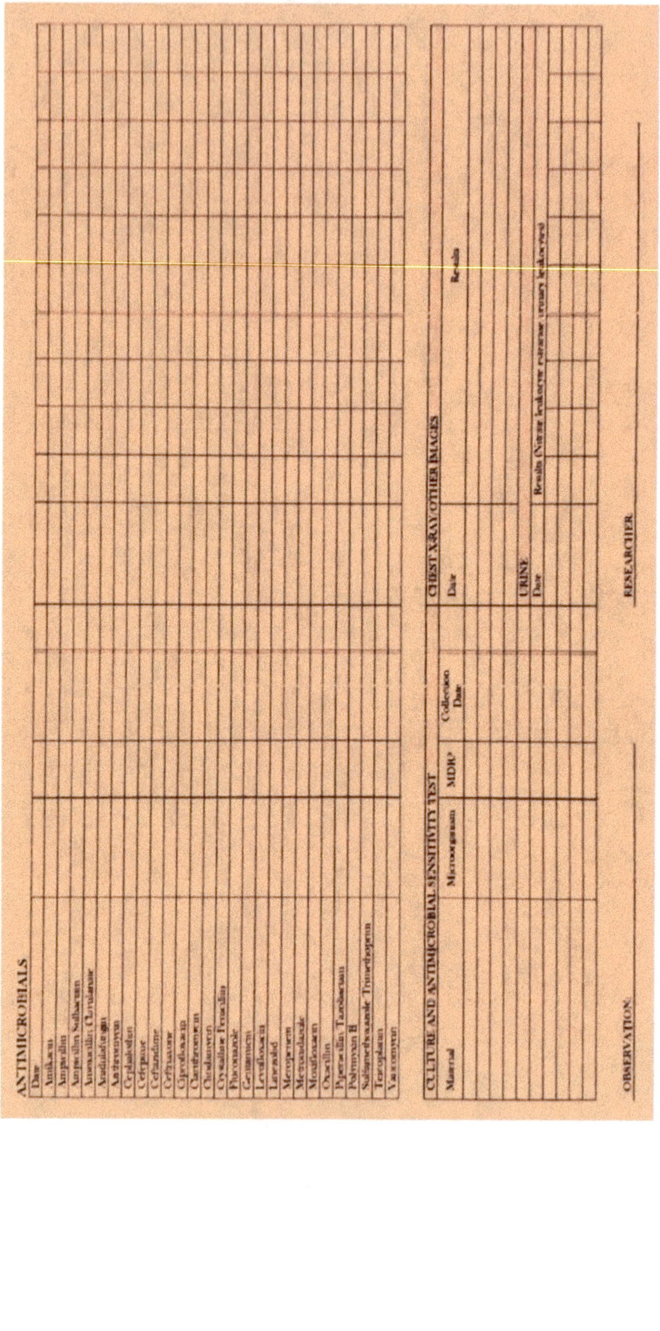

ANTIMICROBIALS

- Day
- Amikacin
- Ampicillin
- Ampicillin Sulbactam
- Amoxicillin Clavulanate
- Azidolate gm
- Aztreonam
- Cefepime
- Cefotaxime
- Ceftazidime
- Ceftriaxone
- Ciprofloxacin
- Clindamycin
- Cefalexin
- Crystalline Penicillin
- Fluconazole
- Gentamicin
- Levofloxacin
- Linezolid
- Meropenem
- Metronidazole
- Moxifloxacin
- Oxacillin
- Piperacillin Tazobactam
- Polymixin B
- Sulfamethoxazole Trimethoprim
- Teicoplanin
- Vancomycin

CULTURE AND ANTIMICROBIAL SENSITIVITY TEST

Material	Macroorganism	MDR?	Collection Date

CHEST X-RAY/OTHER IMAGES

Date	Results

URINE

Date	Results (Nitrate leukocyte esterase urinary leukocytes)

RESEARCHER

OBSERVATION:

Fig 18 – Costs of nosocomial infections – USA – 2013

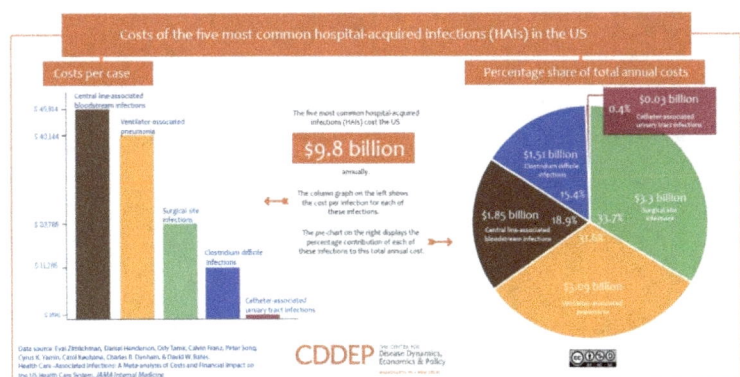

Source: ZIMLICHMAN, HENDERSON, TAMIR, et al., 2013

Figure 19 – Mortality according to the APACHE II score

APACHE II score ranges	Patients (n)	Deaths (n)	Deaths (%)
0 – 5	23	-	-
6 – 10	94	12	12.8
11 – 15	116	24	20.7
16 – 20	134	50	37.3
21 – 25	90	49	54.4
26 – 30	41	28	68.3
30	23	22	95.6
Total	521	185	35.5

Source: CHIAVONE; SENS, 2003. APACHE II score ranges and deaths among 521 Brazilian intensive care unit patients of **Santa Casa de São Paulo** admitted from July 1998 to June 1999